The Class Party

by Amy Ayers

Photographs by Kay McKinley

Developed for Harcourt, Inc., by Gareth Stevens, Inc.
This edition published by Harcourt, Inc., by agreement with Gareth Stevens, Inc. No part of this publication may be reproduced or transmitted in any form or by any means, electronic or mechanical, including photocopy, recording, or any information storage and retrieval system, without permission in writing from the copyright holder.

Requests for permission to make copies of any part of the work should be addressed to Permissions Department, Gareth Stevens, Inc., 330 West Olive Street, Suite 100, Milwaukee, Wisconsin 53212. Fax: 414-332-3567.

HARCOURT and the Harcourt Logo are trademarks of Harcourt, Inc., registered in the United States of America and/or other jurisdictions.

Printed in Mexico

ISBN 13: 978-0-15-360226-9
ISBN 10: 0-15-360226-0

3 4 5 6 7 8 9 10 050 16 15 14 13 12 11 10 09 08

Harcourt
SCHOOL PUBLISHERS

Today is the last day of school. Miss Green's class is having a party. There are muffins to share. There is juice too.

The children will eat together. They will play games. They will hear a story. Then it will be time to leave. It will be a fun day!

7

First the children get their muffins. The muffins will be a good snack. There are 7 banana muffins to eat.

$$7 - 5 = 2$$

Miss Green gives the children 5 banana muffins from the table. How many muffins are left? There are 2 banana muffins left.

8

Now the other children get their muffins.
These muffins have tasty blueberries in them.
There are 8 blueberry muffins to eat.

$$8 - 4 = 4$$

Miss Green gives the children 4 blueberry muffins from the table. How many muffins are left? There are 4 blueberry muffins left.

3

The children choose their juice. There is grape juice. There is apple juice too. There are 3 children who want grape juice to drink.

$$3 + 7 = 10$$

There are 7 more children who want grape juice. There are 10 children in all who want grape juice. It tastes great.

1

The children are almost done getting snacks.
Some children still need juice. There is 1 child
who wants to drink apple juice.

$$1 + 4 = 5$$

There are 4 more children who want apple juice. There are 5 children in all who want to drink apple juice. It feels cold.

9

Snack time is over. Now the children choose something to do. The children want to paint. There are 9 paintbrushes for painting.

9 − 5 = 4

They take 5 paintbrushes. How many paintbrushes are left? There are 4 paintbrushes left on the table. Now the children can make pictures.

4

Some children want to play with puppets.

They will give a puppet show for the class.

They have 4 puppets on a tray.

$$4 - 3 = 1$$

The children take 3 puppets from the tray.
How many puppets are left? There is 1 puppet
left on the tray.

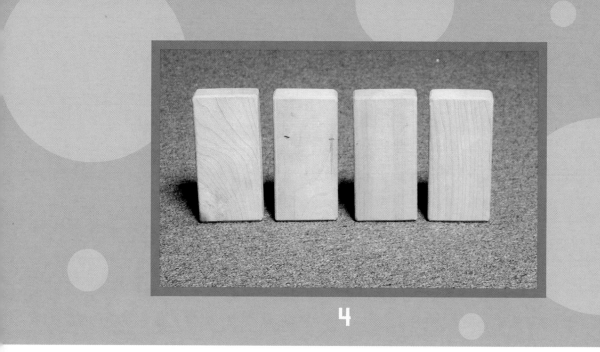

4

Some children want to play with wooden blocks. They will use the blocks to build a house. First they take out 4 big blocks.

4 + 2 = 6

Then the children take out 2 small blocks.
Now they have 6 wooden blocks in all.
They are ready to build the house.

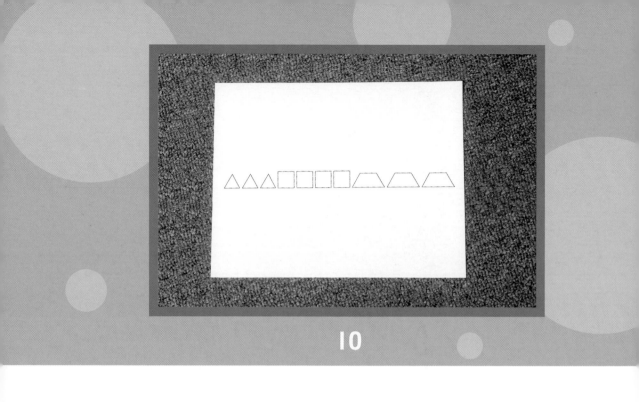

10

Other children want to use blocks to make a pattern. They choose a pattern to make. They need 10 pattern blocks to make this pattern.

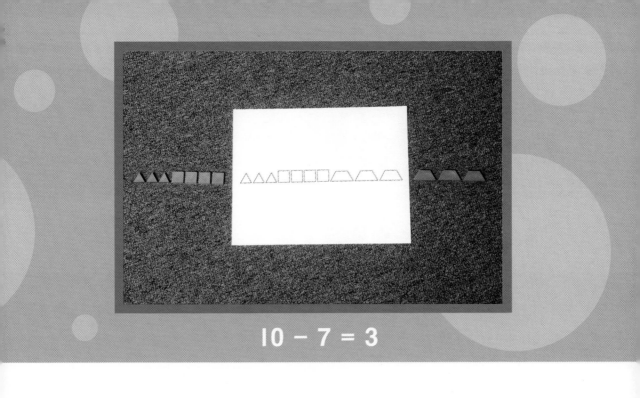

$10 - 7 = 3$

The children have only 7 pattern blocks.
How many more pattern blocks do they need to
make this pattern? They need 3 more
pattern blocks.

9

Soon it is time to clean up. Miss Green will read a story to the class. First 9 children sit down on the rug.

$$9 + 6 = 15$$

Then 6 more children sit down on the rug.

Now 15 children in all are sitting on the rug.

Their teacher reads the story.

The bell rings. It is time to leave. The children say good-bye to Miss Green. She tells them to have a good summer.

It was a fun year for everyone. The children learned a lot. They made many friends.
They cannot wait to come back to school!

Glossary

equals = 1 + 7 = 8
1 plus 7 equals 8.

minus − 8 − 6 = 2
8 minus 6 equals 2.

muffin a small round cake made to serve one person

plus + 6 + 3 = 9
6 plus 3 equals 9.

puppet a doll that fits over the hand